A Differential Equation from a Parallel Universe

A Differential Equation from a Parallel Universe

JULIO CÉSAR MARTÍNEZ ROMERO

Library of Congress Control Number: 2017905633
ISBN: Hardcover   978-1-5065-1997-5
                 Softcover   978-1-5065-1996-8
                 eBook   978-1-5065-1995-1

Print information available on the last page.

Rev. date: 10/04/2017

To order additional copies of this book, contact:
Palibrio
1663 Liberty Drive
Suite 200
Bloomington, IN 47403
Toll Free from the U.S.A 877.407.5847
Toll Free from Mexico 01.800.288.2243
Toll Free from Spain 900.866.949
From other International locations +1.812.671.9757
Fax: 01.812.355.1576
orders@palibrio.com
760741

Contents

ACKNOWLEDGEMENTS

I thank Daniela Carrera Millán, Gilles Figuier, and Noelle Ann Mabry for their kind participation in this book. I also thank and acknowledge the work of Ismael Álvarez León, who kindly took and donated the cover photograph and of Ron Polk for all the other photographs included in the book.

NOTE ON THE ILLUSTRATIONS

An actor's performance in a silent film from the 1920's consisted in a collection of dramatic face expressions. A hundred years ago, those gestures evoked deep emotions in the public. On those days, fourteen-year-old youngsters were thrilled by the films that they watched in a movie theater.

A century of films has changed the definition of visual impact. Nowadays digital images deliver the most awesome experiences on every HD film and videogame. Teens are not easily surprised any more. Their eyes do not stop, not even for a moment, in the 3-D rendering of an image, complex and full of details. The visual environment provokes no thoughts, no inquiries. This book was written for fourteen-year-old youngsters who demand the highest quality of video. This is the reason why I decided that I would not try to fulfill their expectations for image quality and I would resort to an outdated style of pictures, those that moved youngsters and made them dream, reflect, and wonder a hundred years ago.

Hopefully, present day teenagers will look at the illustrations shown here and ask themselves what kind of images they are. Why are these photos here? What is their meaning? What is their relationship to this book?

Maybe the eyes of a youngster will be fixed, at least for a moment, on an illustration. Will it catch a youngster's attention? If teenagers are now used to high-tech digital imagery, let me give them the visual environment that teens of other time would find familiar. Maybe youngsters today will dislike the photographs presented here, but then again, at least they will have noticed them and, in the best scenario, questions might have arisen in their young minds. Questions stir more brain activity than indifference.

INTRODUCTION

There was a time, not long ago, when you were thrilled to spend a Sunday afternoon playing cards with your grandparents. In those times, you loved to go with your granduncle to pet baby animals to the zoo farm on Saturdays. You begged your father to take you to the wild animal park, or to go rowing to the lake.

Nobody expected then, that on a certain day a dormant gene would awaken and it would trigger the production of huge amounts of hormones. These hormones would upset every balance and disturb every aspect of your life. Your brain would be gravely affected. You would be overcome by emotions, thoughts and fears that you would have never imagined before. One day you would be saying the unthinkable and doing the unimaginable. You knew that you were the same child that you used to be, because you shared all his/her history and memories, but his/her likes and dislikes, interests and activities had become strange and unimportant. Now sometimes you feel lost and not even your parents know what to do.

A long time ago, your parents had reinvented their memories of their own younger days. They blocked the uncomfortable episodes, idealized others, and embellished their anecdotes. Therefore, their current interpretation of their own adolescence does not match with their perception of what you are going through. The truth is that they have no clue. They panic because they are facing the unknown. Your parents fear that they are losing control, but it is too late, they have lost all control of what is happening.

Some parents react by denying everything. They avoid confrontation at all cost, become permissive and let you do whatever you want to do. Other parents, realizing that you are trespassing all the limits that were imposed on you when you were a child, increase their authority, provoking you to defy them. There is a third type of parents. There are those who, depending on their mood, are sometimes permissive and other times explode in a burst of anger and fear, and suddenly transform themselves into control freaks. The situation might be exactly the same, their response cannot be more different. Yesterday, your parent was cool with an idea, today the very same idea seems preposterous and offensive.

Teens have a difficult time. On the one hand, you do not understand what is happening to you. You have no idea on what to do with the rollercoaster of your emotions, plans and anxieties. On the other hand, your environment is truly hostile. Your loving parents, whom you admired and to whom you were devoted a couple of years ago, have changed into absurd and sometimes psychotic and abusive strangers.

All your classmates and peers are facing the same problems as you are. Therefore, if there is someone who understands what you are living, it is your friends. You create a link of solidarity with them. They support you and you follow their lead. This bonding makes your parents even more anxious. Your parents used to be your best friends, your confidents. They guided you and you listened to them. Now they lose you to the wild and unwise influence of other disoriented teens. Eventually, the situation becomes unbearable. You prefer to spend the least amount of time possible with your family and to be as long as you can with your friends.

This all happens in the few years between you eleventh birthday and your sixteenth.

And then, there is school with all its boring subjects. The education system pretends that nothing is happening, that you will behave, study, do your homework, prepare your exams, just as you did before. There is so much going around you and none of it has to do with what you are expected to learn at school. On top of everything, there is math.

When math teachers are asked why some students succeed and others fail their subject, their most frequent answer is that the success of the students depends on their talent for math and on the time that they devote to do exercises.

In the very essence of their message, this answer means that those teachers have given up on you. They think, and you share this notion, that you do not have any talent for math. You also find the mathematical exercises so absurd and boring, that there is no way that you are going to spend your weekend studying for an exam, instead of going to the championship game and the parties and after parties that will follow. Thus, the two conditions are fulfilled, you neither have a talent, nor do you have a disposition to study. Therefore, you will inevitably fail.

This book was written for the worst mathematic student ever. Is this you? Did you barely pass your 8th year of elementary education with the lowest promotion grade possible. I wrote it for you, who hate mathematics and do not understand why you have to study what you consider such a boring and absurd subject.

Every mathematics teacher has met students like you. Some teachers think that if they provide you with a lot of demonstrations and formal definitions, you will understand where everything comes from and you will be fascinated by the beauty of abstract logical developments. In my experience, demonstrations have an appeal only for those who already love mathematics.

This book offers the least amount of explanations, everything tries to be clear and mechanical. More than 30 years teaching mathematics have shown me that the first step to transform the attitude towards mathematics is to show the bad students that they can succeed in this subject. Nothing succeeds like success. Later on, you may want to learn the details, or maybe you won't.

The second goal of this book is to show you that equations mean something. They contain information. There is nothing menacing or scary in formulas.

$$\frac{d^2X}{dt^2} - \frac{dX}{dt} - 6X = 0$$

$$\frac{(X-1)^2}{3^2} + \frac{(Y-2)^2}{4^2} = 1$$

Fig. 1

If you find these equations in an exam, do not shiver. You can defeat them easily. You do not need a Ph. D. in nuclear physics to read them and grasp their information content. You only need to know where the clues are. Please give this book a chance and let it show you how. You are only required to read it, page by page, and do whatever it suggests you to do. Please, please, please give this book a chance to transform you.

PART I

Parallel universes

1. Lost in space

This book is about parallel universes, denominated vector spaces, and the passage from one to another. To journey among universes, a transformation device is needed, one that converts a traveler into a parallel being. This device metamorphoses an individual in one vector space to another entity in a different vector space.

We, Charlotte and Jack, are writing from the deck of our spaceship, the Gilgamesh. Our systems failed and we are lost in this infinite vector space, that we call our universe. Our only salvation resides in sending our questions by means of the transformation device to a parallel universe that contains the answers that we need here. Afterwards, we have to receive back the solutions that will let us know our position in our vector space.

Fig. 2

From our vector space, we have been studying a parallel universe that already exists. We watch a different version of ourselves living there, just like a biologist watches a protozoan with a microscope. The protozoan does not know that the biologist exists and it is unaware that it is being observed.

In the universe that bears the secrets that will rescue us, we live a parallel life, oblivious of the knowledge that we possess there and that we cannot discover here. In that vector space, we are ranchers and we raise mammoths. In that universe, we worry about problems that seem inconsequential to us here. Our other selves in that vector space wonder about such trivial things as how to pay our debts or how to have enough room to build new pens for our mammoths, not suspecting that simultaneously, somewhere else, we live a parallel imperiled life.

PART II

A Universe of Ranchers and Mammoths

2. To have or to owe

We are ranchers and we raise mammoths. For our business to prosper, we need to register all our expenses and earnings. We keep a notebook that contains everything related to money, how much we have to pay, how much we earn by selling mammoths, how much we have saved. It also has all the information that concerns the land extensions that we need to raise our mammoths. The third purpose of the notebook is to include a register of mammoths. There we inscribe the dates when the mammoths were born, the names assigned to them, the dates when they were sold or if we kept them for further breeding.

In the end, all the information contained in the book is about numbers, quantities that we add and subtract.

Fig. 3

Take any number, 58 for example. By itself, you cannot tell if it refers to an amount of money, to the area of a piece of land or to the number of mammoths.

Nevertheless, numbers do contain information. There are two kinds of numbers that are unmistakably different. You should never confuse 58 with -58. The former describes what you have, the latter what you owe.

All our money exchanges consist mainly in additions and subtractions of how much money we have and how much money we have to pay.

The gravest mistake we could make would be to forget to write a "-" sign. This small mistake would deceive us into believing that we own the money that we have to pay instead.

If we have to pay $78 for the forage that our mammoths ate last week and $17 for their vitamins, how much do we have to pay in all?

(-78) + (-17) = -95

Thus, when we add our debts, we end up with a larger debt. To know how much, we just have to add the amounts and keep the sign to remember that it is money that we owe. The sign tells us that we do not have the money, we have to pay it.

Now, please help us calculate how much money we owe.

a. (-23) + (-15) =
b. (-34) + (-17) =
c. (-19) + (-13) =

d. (-28) + (-37) =
e. (-43) + (-25) =
f. (-21) + (-35) =
g. (-27) + (-33) =
h. (-23) + (-42) =
i. (-29) + (-44) =
j. (-31) + (-36) =

Answers. a) -38; b) -51; c) -32; d) -65; e) -68; f) -56; g) -60; h) -65; i) -73; j) -67.

If we do have some money and we get more, then we have a larger amount of money. For example,

(78) + (17) = 95

This time the amounts are positive, because we have the money.

Now, please help us calculate how much money we have.

a. (23) + (15) =
b. (34) + (17) =
c. (19) + (13) =
d. (28) + (37) =
e. (43) + (25) =
f. (21) + (35) =
g. (27) + (33) =
h. (23) + (42) =
i. (29) + (44) =
j. (31) + (36) =

Answers. a) 38; b) 51; c) 32; d) 65; e) 68; f) 56; g) 60; h) 65; i) 73; j) 67.

We see that adding numbers with the same sign is easy. If they are positive, we add up the amounts and they keep being positive:

(7) + (8) = 15

If the numbers are negative, it means that we are dealing with debts. We add up the amounts and they keep being negative:

(-7) + (-8) = -15

The difficulty arises when one number is positive and the other is negative. For example, what happens if I have \$15 and I owe \$8?

(15) + (-8)

If I pay my debt, I will still have \$7.

(15) + (-8) = 7

What happens if I have \$8 and I owe \$15?

After I pay as much as I can, I will still owe \$7 and nothing is left for me.

(-15) + (8) = -7

Let us compare both additions and their results. What can we conclude if we add two numbers with different signs?

(15) + (-8) = 7

(-15) + (8) = -7

We realize that to add numbers with different signs, we really have to subtract the smaller from the larger and we keep the sign of the larger.

(15) + (-8) = 7

(-15) + (8) = -7

Now, please help us calculate how much money we owe.

a. $(23) + (-15) =$
b. $(-34) + (17) =$
c. $(19) + (-13) =$
d. $(-28) + (37) =$
e. $(43) + (-25) =$
f. $(21) + (-35) =$
g. $(-27) + (33) =$
h. $(23) + (-42) =$
i. $(-29) + (44) =$
j. $(31) + (-36) =$

Answers. a) 8; b) -17; c) 6; d) 9; e) 18; f) -14; g) 6; h) -19; i) 15; j) -5.

Sometimes we have to add more than two numbers. For example,

forage December	-10
vitamins January	-5
mammoth sold in January	125
veterinary February	-30
forage March	150
mammoth sold in March	-10

To know how much money we have or owe, we separate the positive amounts from the negative ones.

forage December		-10
vitamins January		-5
mammoth sold in January	125	
veterinary February		-30
forage March	150	
mammoth sold in March		-10

In one column we have only the positive numbers, in another the negative ones. One column represents what we have, the other what we owe. Now we add the columns.

forage December		-10
vitamins January		-5
mammoth sold in January	125	
veterinary February		-30
forage March	150	
mammoth sold in March		-10
	275	**-55**

This means that we have $275 and we owe $55.

(275) + (-55) = 220

Now, please help us calculate the following operations.

a. - 2 - 7 + 6 + 4 - 1 =

b. - 1 + 7 - 8 + 9 - 3 =

c. + 6 + 4 + 9 - 3 - 8 =

d. - 5 - 2 - 6 - 4 + 5 =

e. + 9 - 7 + 4 + 6 + 2 =

f. + 3 + 7 - 5 + 4 + 1 =

g. - 6 - 3 + 2 - 8 - 7 =

h. - 4 + 7 - 1 + 5 - 3 =

i. - 3 - 7 + 8 - 4 - 2 =

Answers. a) 0; b) 4; c) 8; d) -12; e) 14; f) 10; g) -22; h) 4; i) -8.

3. Distributing a coefficient

Our friend Wolf MacStoat is our most important trading partner. He buys mammoths from us and he uses them to plough his land. In reciprocity, he sells us the forage that he harvests.

One day, when we had completed our transactions, Wolf MacStoat asked us if we would be interested in tripling our sales. Originally we have added:

-10 - 7 + 120 - 40 - 15

To triple our deal would mean:

3 (-10 - 7 + 120 - 40 - 15)

We just had to multiply everything by 3:

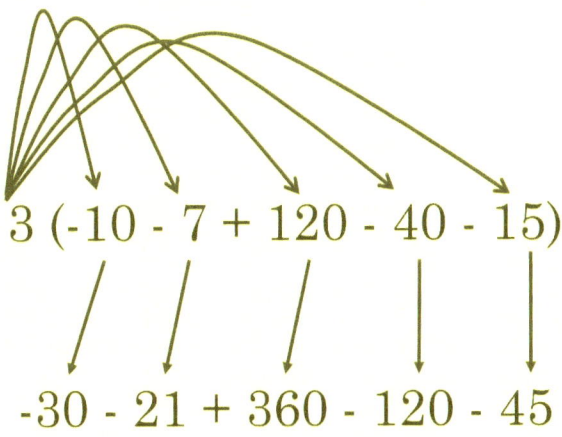

Fig. 4

Now, please help us calculate the following operations.

    a. 3 (- 2 - 7 + 6 + 4 - 1) =
    b. 2 (- 1 + 7 - 8 + 9 - 3) =
    c. 5 (+ 6 + 4 + 9 - 3 - 8) =
    d. 4 (- 5 - 2 - 6 - 4 + 5) =
    e. 6 (+ 9 - 7 + 4 + 6 + 2) =
    f. 3 (+ 3 + 7 - 5 + 4 + 1) =
    g. 7 (- 6 - 3 + 2 - 8 - 7) =
    h. 2 (- 4 + 7 - 1 + 5 - 3) =
    i. 8 (- 3 - 7 + 8 - 4 - 2) =

Answers. a) 0; b) 8; c) 40; d) -48; e) 84; f) 30; g) -154; h) 8; i) -64.

It was just another day when Wolf MacStoat visited our ranch looking for a young mammoth, which he needed to carry logs. None of us noticed that he forgot his accounting notebook on our desk. In that notebook, Wolf MacStoat kept only the business transactions that he does with us. We mistakenly thought that it was our notebook and we did all the arithmetic operations in it, only to realize that it was not our notebook but Wolf's.

Fig. 5

Wolf's

forage April	+25
vitamins May	+10
mammoth sold in May	-100
veterinary May	+30
forage June	+20
mammoth sold in June	-75

+25 +10 - 100 + 30 +20 - 75
Ours:

forage April	-25
vitamins May	-10
mammoth sold in May	+100
veterinary May	-30
forage June	-20
mammoth sold in June	+75

-25 - 10 + 100 - 30 - 20 + 75

Our mistake was not really a problem because it was exactly the same account, except that all the signs were inverted. It was just as is we had multiplied everything by a - sign, like this:

- (+25 +10 - 100 + 30 +20 - 75)

Now we just distribute the sign:

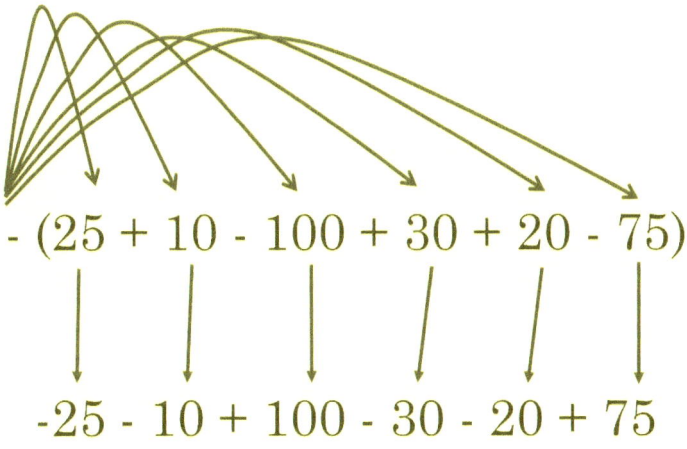

$$- (25 + 10 - 100 + 30 + 20 - 75)$$

$$- (25 + 10 - 100 + 30 + 20 - 75)$$

$$-25 - 10 + 100 - 30 - 20 + 75$$

Fig. 6

We group the negative and the positive:
- 25 - 10 - 30 - 20 + 100 + 75
Add separately the positive and the negative:
- 85 + 175 = 90
Now, please help us calculate the following operations.

a. - (- 2 - 7 + 6 + 4 - 1) =
b. - (- 1 + 7 - 8 + 9 - 3) =
c. - (+ 6 + 4 + 9 - 3 - 8) =
d. - (- 5 - 2 - 6 - 4 + 5) =
e. - (+ 9 - 7 + 4 + 6 + 2) =
f. - (+ 3 + 7 - 5 + 4 + 1) =
g. - (- 6 - 3 + 2 - 8 - 7) =
h. - (- 4 + 7 - 1 + 5 - 3) =
i. - (- 3 - 7 + 8 - 4 - 2) =

Answers. a) 0; b) -4; c) -8; d) 12; e) -14; f) -10; g) 22; h) -4; i) 8.

We are now ready to multiply by a negative number:

The minus sign changes the signs of everything inside the parenthesis. Everything positive becomes negative and vice versa.

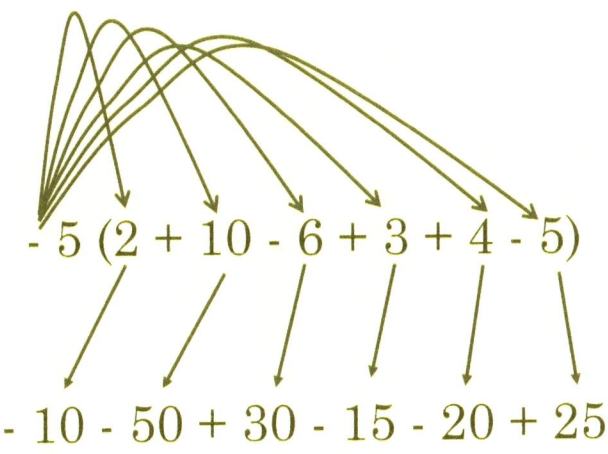

Fig. 7

- 10 - 50 + 30 - 15 - 20 + 25

We group the negative and the positive:

- 10 - 50 - 15 - 20 + 30 + 25

Add separately the positive and the negative:

- 95 + 55 = - 40

Now, please help us calculate the following operations.

 a. - 2 (- 2 - 7 + 6 + 4 - 1) =
 b. - 3 (- 1 + 7 - 8 + 9 - 3) =
 c. - 5 (+ 6 + 4 + 9 - 3 - 8) =
 d. - 2 (- 5 - 2 - 6 - 4 + 5) =
 e. - 4 (+ 9 - 7 + 4 + 6 + 2) =
 f.   - 3 (+ 3 + 7 - 5 + 4 + 1) =
 g. - 5 (- 6 - 3 + 2 - 8 - 7) =
 h. - 2 (- 4 + 7 - 1 + 5 - 3) =
 i.   - 3 (- 3 - 7 + 8 - 4 - 2) =

Answers. a) 0; b) -12; c) -40; d) 24; e) -56; f) -30; g) 110; h) -8; i) 24.

If we want to reduce our business to half the one we previously intended, we just divide everything by two:

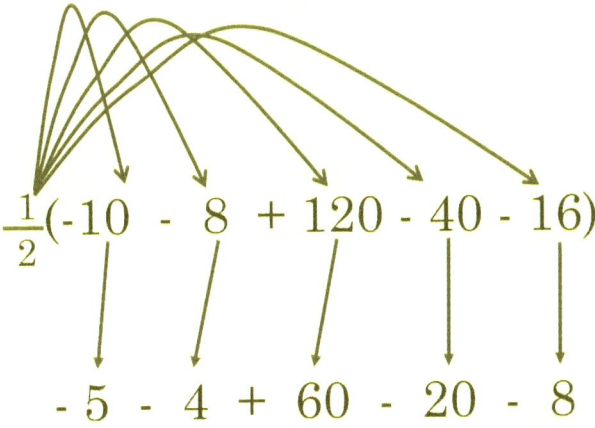

$$\frac{1}{2}(\text{-}10 - 8 + 120 - 40 - 16)$$

$$\frac{1}{2}(\text{-}10 - 8 + 120 - 40 - 16)$$

$$- 5 - 4 + 60 - 20 - 8$$

Fig. 8

- 5 - 4 + 60 - 20 - 8

We group the negative and the positive:

- 5 - 4 - 20 - 8 + 60

Add separately the positive and the negative:

- 37 + 60 = 23

Now, please help us calculate the following operations.

a.   - ½ (- 2 - 7 + 6 + 4 - 1) =
b.   - ½ (- 1 + 7 - 8 + 9 - 3) =
c.   - ½ (+ 6 + 4 + 9 - 3 - 8) =
d.   - ½ (- 5 - 2 - 6 - 4 + 5) =
e.   - ½ (+ 9 - 7 + 4 + 6 + 2) =
f.   - ½ (+ 3 + 7 - 5 + 4 + 1) =
g.   - ½ (- 6 - 3 + 2 - 8 - 7) =
h.   - ½ (- 4 + 7 - 1 + 5 - 3) =
i.   - ½ (- 3 - 7 + 8 - 4 - 2) =

Answers. a) 0; b) -2; c) -4; d) 6; e) -7; f) -5; g) 11; h) -2; i) 4.

4. The signs of products and divisions

We have been dealing with numbers that are attached to a sign.

+ 5 means that we have 5.

- 5 means that we owe 5.

A positive (+) sign can be seen as true and as negative sign (-) as false.

Therefore, + means "it is true" and - means "it is false".

Let us consider a false (-) statement.

Mammoths can fly.

It is false (-).

What about the statement?

It is false (-) that mammoths can fly (-). This statement is true (+).

Thus, (-)(-) = +

This other statement contains different information.

It is true (+) that mammoths can fly (-). This statement is false (-).

(+)(-) = -

Now let us consider a true (+) statement.

Falcons can fly (+).

It is true (+) that falcons can fly (+). This statement is true (+).

(+)(+) = +

It is false (-) that falcons can fly (+). This statement is false (-).

(-)(+) = -

Fig. 9

Now let us summarize all these data in a table.

-	-	=	+
+	-	=	-
+	+	=	+
-	+	=	-

The analysis of the table leads us to conclude that two equal signs give us a + and two different signs give us a minus sign.

Now, with these ideas in mind let us multiply the following pairs of numbers.

(-3)(-2) = +6, equal signs give a +.

(-3)(+2) = -6, different signs give a -.

(+3)(+2) = +6, equal signs give a +.

(+3)(-2) = -6, different signs give a -.

These results lead us to a conclusion. When multiplied, each pair of minus signs becomes a +. An odd number of minus signs will still be a minus.

Examples.

$$(-3)(-5)(2)(5)(-1)(-2)$$

$$\uparrow \quad \uparrow \qquad \qquad \uparrow \quad \uparrow$$

an even number of minus signs
becomes a +

$$+300$$

$$(3)(5)(-2)(-5)(-1)(2)$$

$$\uparrow \quad \uparrow \quad \uparrow$$

an odd number of minus signs
keeps the minus sign

$$-300$$

$$(-3)(-5)(-2)(-5)(-1)(-2)$$

$$\uparrow \quad \uparrow \quad \uparrow \quad \uparrow \quad \uparrow \quad \uparrow$$

an even number of minus signs
becomes a +

$$+300$$

Fig. 10

Therefore, when multiplying several numbers with signs, we only need to count the number of minus signs and then follow the rule:

An even number of minus becomes a +.

An odd number of minus keeps the minus sign.

Now, let us practice the new skills that we have just acquired.

If we multiply,

(4)(5)(2)(5)(6)(5)

The result is 6000.

Now let us multiply numbers with signs.

a. (-4)(5)(-2)(-5)(-6)(5)
b. (4)(5)(-2)(-5)(6)(5)
c. (-4)(5)(2)(-5)(6)(-5)
d. (4)(-5)(2)(5)(-6)(-5)
e. (4)(-5)(2)(5)(-6)(5)
f. (-4)(-5)(-2)(5)(-6)(-5)

Answers. a) 6000; b) 6000; c) -6000; d) -6000; e) 6000; f) -6000.

Now we will go up to the next level of complexity. We will multiply two additions of numbers, (+3) + (-2) and (+6) + (-4) + (-2).

((+3) + (-2)) ((+6) + (-4) + (-2))

((+3) + (-2)) ((+6) + (-4) + (-2))

(+18) + (-12) + (-6)

((+3) + (-2)) ((+6) + (-4) + (-2))

(-12) + (+8) + (+4)

(+18) + (-12) + (-6) + (-12) + (+8) + (+4)
We group the positive and the
negative numbers.
(+18 + 8 + 4) + (-12 -6 - 12)
(+30) + (-30) = 0

Fig. 11

Let us practice this new skill.

a. $(+2 - 1)(1 - 2 - 3) =$
b. $(+2 - 1)(-1 + 2 + 3) =$
c. $(+2 + 1)(-1 + 2 - 3) =$
d. $(+2 + 1)(1 - 2 + 3) =$
e. $(-2 + 1)(1 + 2 - 3) =$
f. $(-2 + 1)(-1 - 2 + 3) =$
g. $(-2 - 1)(-1 - 2 - 3) =$
h. $(-2 - 1)(1 + 2 + 3) =$

Answers. a) -4; b) 4; c) -6; d) 6; e) 0; f) 0; g) 18; h) -18.

PART III

An interference.
A universe of ocelots

5. Two cubs kidnapped, a monkey and an ocelot

I am writing from the deck of our spaceship. We had been closely watching a parallel universe of ranchers and mammoths because that universe holds the answer to our problems.

Fig. 12

Suddenly, interference blocked the signal and we started receiving images from another vector space, a universe of ocelots. In that universe, there was a family of ocelots. The family consisted of a mother, Mama Coty, her little brother, Felix, and her three daughters, Esperanza, Tere and Clarita. They lived in the jungle.

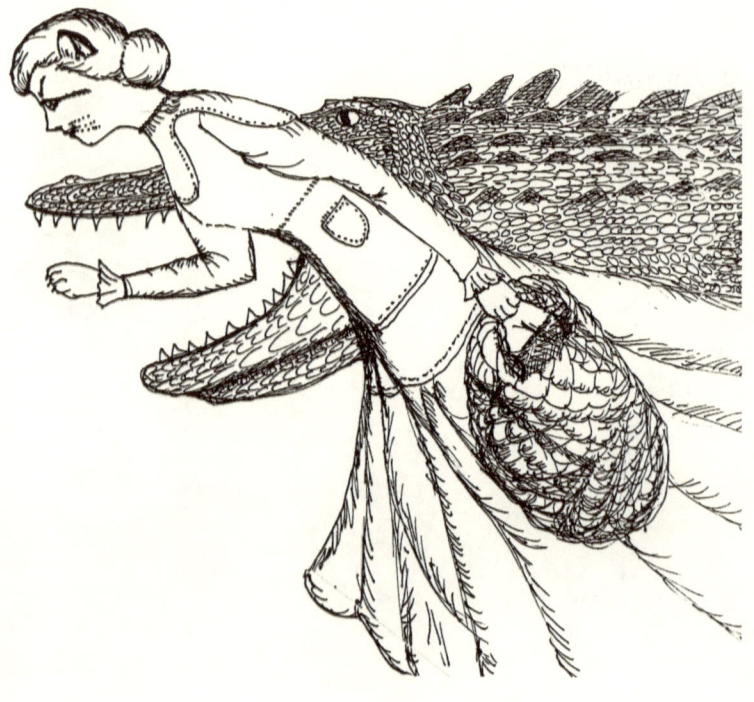

Fig. 13

The level of insecurity in their country was so unbearably high, that Mama Coty had taken her little brother and her three daughters to a border town with the purpose of crossing to another country. The ocelots spent the night in an inn. Out on the streets, there were gangs of squirrel monkeys all around.

Everyone was talking about a terrible event that had recently happened. The Prince of the squirrel monkeys had been kidnapped by traffickers and was sold to a wealthy human businessman. The King of the squirrel monkeys had contacted the businessman, hoping to negotiate the release of his son. The businessman agreed to liberate the squirrel monkey prince in exchange for another exotic pet. The King ordered his squirrel monkey soldiers to bring him any cute cub that could be a replacement for his son as a pet. Therefore, gangs of squirrel monkeys had invaded all small towns looking for a suitable substitute for the prince. When Mama Coty and her family arrived to the border town, this was the topic of every conversation.

As soon as they arrived, Mama Coty gave each ocelot in her family a roasted quail for the next stage of their trip. The following morning they woke up and started their journey to the mountains. Mama Coty had been told that there were magnificent meadows at the other side of the mountains, safe from danger and full of quarry.

Mama Coty set up her mind to go there to establish themselves.

They had walked for about an hour when Clarita realized that she had forgotten her roasted quail in the hotel room. Without telling anyone where she was going, Clarita ran back to the inn.

At the door of the inn, Clarita paused for a while to catch her breath and walked in. There was a troop of squirrel monkeys at the hall. They all watched her with interest and exchanged weird comments that she could not understand. When Clarita came back from the bedroom with her roasted quail, the squirrel monkeys started teasing her and hindered her exit. She did not know what to do and was very scared.

Fig. 14

Not long after Clarita had left the ocelot family, the rest of the party noticed her absence. Her mother, sisters and uncle looked for her and called her name hoping that she would answer. Sadly, there was no hint of her whereabouts. After pondering on the situation, the four of them traced back their steps and returned to the village. Nowhere could Clarita be found.

Fig. 15

They walked back to the inn and asked if anyone had seen the girl. They soon learned what had happened. The gang of squirrel monkeys had kidnapped her and had taken her to their headquarters in the middle of the jungle. The squirrel monkeys hoped that Clarita would be a suitable substitute and that she would help them recover their Prince.

6. Crossing the swamp

To get back to the jungle, Mama Coty and her family had to cross the swamps. Mama Coty feared that it would take them too long to find Clarita. She was also afraid that the squirrel monkeys might have closed their deal with the humans and had exchanged Clarita for their Prince. Therefore, Mama Coty asked a manatee to let them ride on its back in order to cross the swamps as quickly as possible.

- Dear manatee, we need to cross the swamps as fast as possible so that I can rescue my daughter. Please let us ride on your back to the border of the jungle. - asked Mama Coty.

- I will gladly do as you request. Nevertheless, I have something to ask in return. My crown used to have ten pearls on it. One day some fish stole one pearl each and only three are left. If you bring me back my pearls, I will let you ride on my back through the streams to the border of the jungle.

X pearls stolen +   3 left in the crown= total

X   +   3   = 10

Fig. 16

42 JULIO CÉSAR MARTÍNEZ ROMERO

Answer X = 7.

Mama Coty went to look for the fish and explained to them why she needed their help.

- Dear fish, please give me back the pearls that you took from the manatee's crown so that it will let us ride on its back across the swamps as fast as possible to the border of the jungle so that I can rescue my daughter. - asked Mama Coty.

- We will gladly do as you request. Nevertheless, we have something to ask in return. The branches of a mango tree hang above the stream. The fruit used to fall in the stream, attracting insects that we ate. One day, three macaws took our mangoes and now we have no insects to eat. If you bring us back the mangoes that a single macaw took, there will be enough insects for us to eat. Then we will gladly give you the pearls that you need.

- How many mangoes did each macaw take?

- In all there were 18 mangoes, three fell in the forest and the others were equally distributed among the three macaws.

3 macaws with   3 mangoes fell
X mangoes each   in the forest =   total

3X   +   3   =   18

Fig. 17

Answer X = 5.

Mama Coty went to look for the macaws and explained to them why she needed their help.

- Dear macaws, please give me the mangoes that one of you have, so that the fish will give me back the pearls that they took from the manatee's crown and then it will let us ride on its back across the swamps as fast as possible to the border of the jungle so that I can rescue my daughter. - asked Mama Coty.

- We will gladly do as you request. Nevertheless, we have something to ask in return. Please ask the raccoon to give us the shrimps that he catches in the stream on one day. - The macaws replied.

- How many shrimps are those?

- If you go to see him right now, he will be catching shrimps and he will inform you. - The macaws said.

Mama Coty and her family went to look for the raccoon. He was very busy.

- Instead of standing there just watching me, why don't you help me catch shrimps? - The raccoon said when he saw them.

Each of the ocelots caught a shrimp.

- How many shrimps do you catch each day? - Mama Coty asked him.

- In three days, I will have 34, including the 4 that you just caught.

day 1   day 2   day 3

$$x + x + x + 4 = 34$$

3 days collecting+   one shrimp collected
   shrimps   by each ocelot = total

   3X   +   4   =   34

Fig. 18

Answer X = 10.

- Dear Mr. Raccoon, please give me the shrimps that you catch on a day. I will give them to the macaws. In turn, one of them will give me its mangoes. In exchange of the mangoes, the fish will give me back the pearls that they took from the manatee's crown and then it will let us ride on its back across the swamps as fast as possible to the border of the jungle so that I can rescue my daughter. - asked Mama Coty.

- I will gladly do as you request. In return, please ask one of the four squirrels to give me back the hazelnuts that it took away from me. - The raccoon replied.

- How many hazelnuts does each squirrel have?

- The four squirrels have 90 hazelnuts in all, including 10 in a basket. The rest of the hazelnuts were distributed equally among the four squirrels.

$$X + X + X + X + 10 = 90$$

4 squirrels   10 hazelnuts in
each with X hazelnuts +   a basket   = total

4X   +   10   =   90

Fig. 19

Answer X = 20.

Mama Coty and her family went to look for the squirrels.

When the ocelots found the squirrels, Mama Coty asked them.

- Will you please give me the hazelnuts that any of you have? I will give them to the raccoon. He will give me the shrimps that he catches on a day. I will give the shrimps to the macaws. In turn, one of them will give me its mangoes. In exchange of the mangoes, the fish will give me back the pearls that they took from the manatee's crown and then it will let us ride on its back across the swamps as fast as possible to the border of the jungle so that I can rescue my daughter. - asked Mama Coty.

-We will give you nothing. The hazelnuts are ours and we will not share them. - Was their answer.

The four ocelots jumped and each one of them caught a squirrel.

- We, ocelots, eat squirrels. If you don't give us the hazelnuts, we will eat you right away. - Mama Coty warned them.

Thus, the squirrels gave them the hazelnuts.

Fig. 20

As soon as the ocelots had the hazelnuts, Mama Coty gave them to the raccoon. He gave her the shrimps that he caught on a day. Mama Coty gave the shrimps to the macaws. In turn, one of them gave her its mangoes. In exchange of the mangoes, the fish gave Mama Coty the pearls that they had taken from the manatee's crown. Finally, the manatee let the ocelots ride on its back across the swamps as fast as possible to the border of the jungle where Mama Coty hoped to rescue Clarita, her daughter.

Fig. 21

7. Travelling through the jungle

Once the ocelots were in the jungle, they asked a herd of brocket deer if they had seen a troop of squirrel monkeys.

- Indeed, we saw their darks shapes on the canvas of the trees above us. Indeed, we heard their screaming and shouting as they noisily proceeded on their journey. Indeed, they were carrying a prisoner with them. - The brocket deer replied.

- Do you know where they were going? - Mama Coty asked.

- Indeed, they were heading to meet the human poachers. Humans have taken their Prince and the monkeys give them every young cute creature they kidnap, hoping to rescue their Prince. That is the sad truth, indeed. - Was the deer's answer.

- The squirrel monkeys stole my daughter. Please let us ride on your backs so that we can get to their meeting place before my daughter is delivered to the human poachers. - begged Mama Coty.

- We will gladly do as you request. Nevertheless, we have something to ask in return. Five anteaters have been gathering quartz crystals. They each have the same amount. We want to own the crystals that any of them would give us. - The deer explained.

- How many crystals are those?

- In all, they have 50 crystals. That number includes 10 that they will bring to their King. The rest, they have divided equally among the five of them.

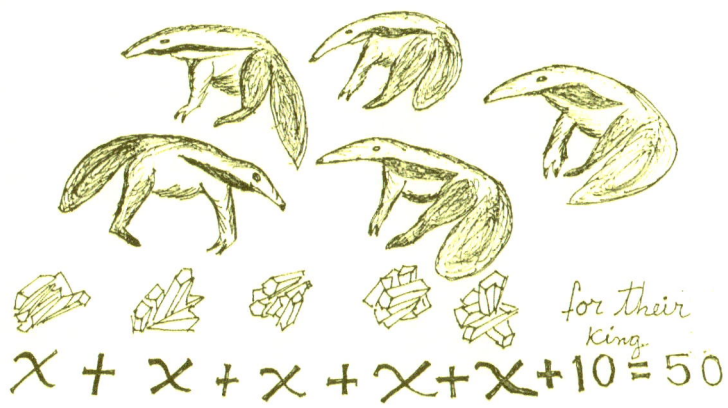

X + X + X + X + X + 10 = 50

5 anteaters
with X crystals each +

10 crystals for
their king   **total**

5X   +   10   =   50

Fig. 22

Answer X = 8.

Mama Coty looked for the anteaters.

- Will any among you please give me your quartz crystals? The squirrel monkeys stole my daughter. In exchange for the crystals, four brocket deer will let us ride on their backs so that we can get to the squirrel monkeys' meeting place before my daughter is delivered to the human poachers.

- I will gladly give you my quartz crystals, - One of the anteaters said. - but I will ask you for something in return. The river otter collects beautiful shells. She has three boxes, two of them full with the same amount of shells. The third box has not been completed yet, it still lacks six shells to be full. In all, the river otter has 30 shells. I will exchange my quartz crystals for a box full of shells.

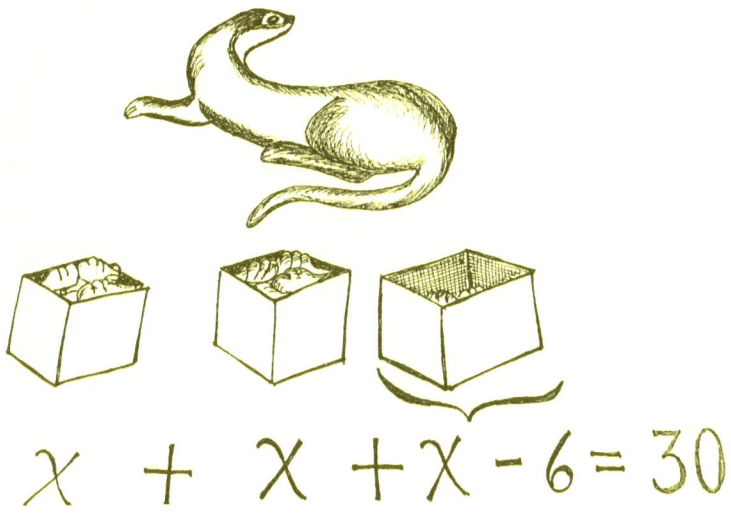

$$\chi + \chi + \chi - 6 = 30$$

3 boxes with   6 are still needed
X shells in each box -   to complete a box   total

3X   -   6   = 30

Fig. 23

Answer X = 12.

Mama Coty went to look for the river otter.

- Will you please give me a box full of shells? An anteater will accept the shells in exchange for its quartz crystals. In exchange for the crystals, four brocket deer will let us ride on their backs so that we can get to the squirrel monkeys' meeting place. The squirrel monkeys stole my daughter. I hope to arrive on time before my daughter is delivered to the human poachers.

- I will gladly give you the shells but I must ask for something in return. The forest snake is the guardian of an ancient Spanish treasure. There are five chests with gold coins and there are other five coins in a small velvet bag. In all, there are 65 coins. If you bring to me the coins of one of the chests, I will gladly give you the shells that you require.

$$x + x + x + x + x + 5 = 65$$

5 chests of coins + one bag of coins = total

5X   +   5   = 65

Fig. 24

Answer X = 12.

Mama Coty went to look for the snake.

- Will you please give me a chest of gold coins? The river otter asked me for those coins in exchange for a box of shells. An anteater will accept the shells in exchange for its quartz crystals. In exchange for the crystals, four brocket deer will let us ride on their backs so that we can get to the squirrel monkeys' meeting place. The squirrel monkeys stole my daughter. I hope to arrive on time before my daughter is delivered to the human poachers.

- I will gladly give you the coins but I must ask for something in return. The opossum collects sapphires. He has ten bags of the blue stones. Nine bags have the same amount of sapphires but one bag is incomplete. Five sapphires would be needed to complete it. In all, the opossum has 115 sapphires. If you bring to me the sapphires of one of the full bags, I will gladly give you the coins that you require.

$$\chi + \chi + \chi + \chi + \chi + \chi + \chi + \chi + \chi + \chi - 5 = 115$$

10 bags with sapphires	-	5 needed to complete a bag	=	total
10X	-	5	=	115

Fig. 25

60 JULIO CÉSAR MARTÍNEZ ROMERO

Answer X = 12.

Mama Coty went to look for the opossum.

- Will you please give me a bag full of sapphires? In exchange for them the snake will give me a chest full of gold coins. The river otter asked me for those coins in exchange for a box of shells. An anteater will accept the shells in exchange for its quartz crystals. In exchange for the crystals, four brocket deer will let us ride on their backs so that we can get to the squirrel monkeys' meeting place. The squirrel monkeys stole my daughter. I hope to arrive on time before my daughter is delivered to the human poachers.

- I will gladly give you the sapphires but I must ask for something in return. There is a coati in the jungle. He has three nets full of tangerines and seven tangerines in a tray. In all, the coati has 31 tangerines. If you bring to me the tangerines contained in one of the nets, I will gladly give you the sapphires that you require.

$$x + x + x + 7 = 31$$

3 nets with X tangerines each	+	7 tangerines in a tray	= total
3X	+	7	= 31

Fig. 26

Answer X = 8.

Mama Coty went to look for the coati.

- Will you please give me a net full with tangerines? The opossum wants them to trade for a bag full of sapphires. In exchange for them the snake will give me a chest full of gold coins. The river otter asked me for those coins in exchange for a box of shells. An anteater will accept the shells in exchange for its quartz crystals. In exchange for the crystals, four brocket deer will let us ride on their backs so that we can get to the squirrel monkeys' meeting place. The squirrel monkeys stole my daughter. I hope to arrive on time before my daughter is delivered to the human poachers.

- I will gladly give you the tangerines but I must ask for something in return. The wild turkey stole my wife and he does not want to give her back to me. If you bring back my wife to me, I will gladly give you the sapphires that you require.

Fig. 27

Mama Coty went to look for the wild turkey.

- Will you please let the wife of the coati go free? If you let her free, the coati will give me a net full with tangerines. The opossum wants the tangerines to trade for a bag full of sapphires. In exchange for them the snake will give me a chest full of gold coins. The river otter wants those coins in exchange for a box of shells. An anteater will accept the shells in exchange for its quartz crystals. In exchange for the crystals, four brocket deer will let us ride on their backs so that we can get to the squirrel monkeys' meeting place. The squirrel monkeys stole my daughter. I hope to arrive on time before my daughter is delivered to the human poachers.

- I will never let the wife of the coati go free. - Replied the wild turkey.

- I am an ocelot. Ocelots eat wild turkeys. If you do not release the coati's wife, my family and I will eat you for lunch. - Mama Coty warned the turkey.

Fig. 28

Thus, the wild turkey released the coati's wife. The coati was so happy to see his wife again that he gladly gave Mama Coty a net full with tangerines. Once he had the tangerines, the opossum gave the ocelots a bag full of sapphires. In exchange for them, the snake gave Mama Coty a chest full of gold coins. The river otter took the coins and gave the ocelots a box of shells. An anteater received the shells in exchange for its quartz crystals. In exchange for the crystals, four brocket deer let the four ocelots ride on their backs to the squirrel monkeys' palace.

Fig. 29

8. Rescuing a monkey and an ocelot by teaching algebra

The troop of squirrel monkeys was already in the Capital City of their Kingdom. The ocelots arrived there on the backs of the brocket deer at the same time as the poachers got there. The poachers and the ocelots entered the royal palace simultaneously. Mama Coty was the first to speak.

- Honorable King of the Squirrel Monkeys, your soldiers have taken my daughter to offer her to these poachers who have also come to meet you.

- You are right. - The King replied - This is a rather unfortunate situation. I will do anything to recover my son, who is kept by the humans.

- There might be a solution for this conundrum.
- The chief of the poachers addressed the King. - Noble King, your son is kept as the pet of a young boy, the son of a wealthy businessman. The boy spends too much time playing with the monkey and neglects his studies. His father is very angry with your son for this reason. I am sure that the businessman will gladly give you back your son, the Prince, if you help him.

- How can I help him? - Mama Coty and the King asked simultaneously.

- The boy needs some tutoring in algebra.

Among the ocelot teens, Tere and Esperanza were the brightest of the students and excelled in their grades in every subject.

- We will help the human student learn algebra so that he releases the Squirrel Monkey Prince but you must give us our sister back. - Tere and Esperanza said.

- I will keep Clarita as leverage until my son is sound and safe back home. - The King pronounced.

Clarita was kept as a hostage of the King of the squirrel monkeys.

Fig. 30

Mama Coty remained in the squirrel monkey city with Clarita. Tere, Esperanza and Felix travelled with the poachers to the human town to teach algebra to the businessman's son.

The poachers took the ocelot girls to the businessman's house. The businessman explained the situation to them.

- If my son does not hand in this algebra homework, he will fail the course and he will not be admitted in high school.

- I need to learn how to solve linear equations. - The human student explained to the ocelot girls.

Fig. 31

9. The need to keep the balance

It was Tere who started the explanation.

- In a chest, there is an unknown amount of gold coins.

- How many? We do not know. - Felix said.

- Ten, one hundred, twelve? Who knows? - Esperanza wondered.

- We call that amount X. There are X coins in the chest.

- An equation describes the situation involved in the problem.

- There are three chests with the same amount of coins each (3X). Additionally, there is a small velvet bat with 14 coins. In all there are 50 coins.

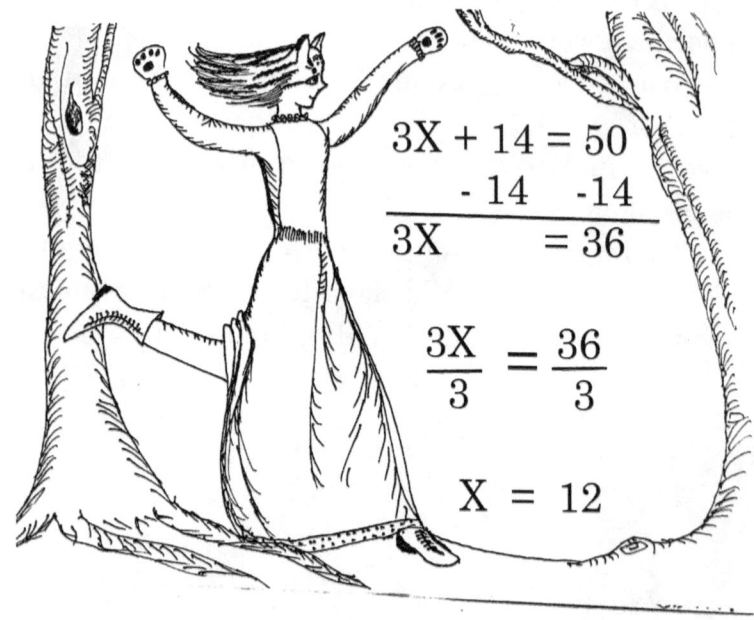

$$3X + 14 = 50$$
$$\underline{\quad -14 \quad\quad -14 \quad}$$
$$3X \quad\quad\quad = 36$$

$$\frac{3X}{3} = \frac{36}{3}$$

$$X = 12$$

Fig. 32

- To solve an equation means to find the value of X.

- The exact number.

- To solve an equation, we must pay attention to the equal sign (=).

- The equal sign divides the equation.

- It separates the left from the right.

- There is the same number of coins on the left side, as there are on the right.

- It is like placing weights on a scale. To be balanced, each plate must have exactly the same weight.

Fig. 33

- To keep the balance, whatever we do on one side, we have to do on the other side.

$$4X + 6 = 50$$

If we subtract six on one side, we have to subtract six on the other.

$$
\begin{array}{l}
4X + 6 = 50 \\
 -6 \quad -6 \\
\hline
4X = 44
\end{array}
$$

If we divide by four one side, we have to divide by four the other.

$$\frac{4X}{4} = \frac{44}{4}$$

$$X = 11$$

Fig. 34

We keep operating on both sides of the equation, until we find the value of X.

Let us review some more examples.

$$\tfrac{1}{2}X + 5 = 25$$

If we subtract five on one side, we have to subtract five on the other.

$$
\begin{array}{rl}
\tfrac{1}{2}X + 5 &= 25 \\
-5 & \quad -5 \\
\hline
\tfrac{1}{2}X &= 20
\end{array}
$$

If we multiply by two one side, we have to multiply by two the other.

$$2\,(\tfrac{1}{2}X) \quad = \quad 2(20)$$

$$X \quad = \quad 40$$

Fig. 35

$$4X + 2 = X + 32$$

If we subtract two on one side, we have to subtract two on the other.

$$
\begin{array}{l}
4X + 2 = X + 32 \\
\quad\ - 2 \qquad\quad -2 \\
\hline
4X \qquad = X + 30
\end{array}
$$

If we subtract X on one side, we have to subtract X on the other.

$$
\begin{array}{l}
4X \qquad = X + 30 \\
-X \qquad\quad -X \\
\hline
3X \qquad = \quad + 30
\end{array}
$$

If we divide by three one side, we have to divide by three the other.

$$\frac{3X}{3} = \frac{30}{3}$$

$$X = 10$$

Fig. 36

$$7X - 5 = X + 25$$

If we add five on one side, we have to add five on the other.

$$7X - 5 = X + 25$$
$$\underline{ + 5 + 5}$$
$$7X = X + 30$$

If we subtract X on one side, we have to subtract X on the other.

$$7X = X + 30$$
$$\underline{-X -X}$$
$$6X = + 30$$

If we divide by six one side, we have to divide by six the other.

$$\frac{6X}{6} = \frac{30}{6}$$

$$X = 5$$

Fig. 37

$5X - 2 = 34 - X$

If we add two on one side, we have to add two on the other.

$$\begin{array}{rcl} 5X - 2 &=& 34 - X \\ + 2 & & +2 \\ \hline 5X &=& 36 - X \end{array}$$

If we add X on one side, we have to add X on the other.

$$\begin{array}{rcl} 5X &=& 36 - X \\ +X & & +X \\ \hline 6X &=& 36 \end{array}$$

If we divide by six one side, we have to divide by six the other.

$$\frac{6X}{6} = \frac{36}{6}$$

$X = 6$

Fig. 38

$8X - 3 = 30 - 3X$

If we add three on one side, we have to add three on the other.

$$8X - 3 = 30 - 3X$$
$$\underline{+ 3 \quad +3 }$$
$$8X \quad\ = 33 - 3X$$

If we add 3X on one side, we have to add 3X on the other.

$$8X \quad\ = 33 - 3X$$
$$\underline{+3X +3X}$$
$$11X \quad = 33$$

If we divide by 11 one side, we have to divide by 11 the other.

$$\frac{11X}{11} = \frac{33}{11}$$

$$X = 3$$

Fig. 39

-5X + 8 = 7X - 40

If we add 40 on one side, we have to add 40 on the other.

$$
\begin{array}{rcl}
-5X + 8 & = & 7X - 40 \\
+ 40 & & +40 \\
\hline
-5X + 48 & = & 7X
\end{array}
$$

If we add 5X on one side, we have to add 5X on the other.

$$
\begin{array}{rcl}
-5X + 48 & = & 7X \\
+5X & & + 5X \\
\hline
48 & = & 12X
\end{array}
$$

If we divide by 12 one side, we have to divide by 12 the other.

$$
\frac{48}{12} = \frac{12X}{12}
$$

4 = X, that is X = 4

Fig. 40

Now let us practice the new skills that we have acquired.

a. $4X - 5 = 15 - 6X$
b. $6X + 8 = 4X + 16$
c. $3X + 4 = 2X + 8$
d. $8X - 14 = 26 - 12X$
e. $-3X + 4 = 2X - 6$
f. $-3X + 4 = -2X + 6$
g. $-3X - 4 = 2X + 6$
h. $3X + 4 = -2X + 14$
i. $3X - 4 = 2X - 6$
j. $3X + 4 = -2X - 6$

Answers. a) $X = 2$; b) $X = 4$; c) $X = 4$; d) $X = 2$; e) $X = 2$; f) $X = -2$; g) $X = -2$; h) $X = 2$; i) $X = -2$; j) $X = -2$.

PART IV

Back to the universe of ranchers and mammoths

10. Trapezoids and the meaning of X^2

Young mammoths are very fragile creatures. Their mothers usually take good care of them. However, every once in a while, a mother is too young and does not know what to do with her calf. Whenever we need to isolate a mammoth, we ask our friend Wolf MacStoat, to build an adequate pen for that specific mammoth.

Fig. 41

Today we need a 4m x 5m pen for a baby mammoth. The pen consists of 20 squares, each side measuring 1m. Thus, the area of the pen is 20 square meters.

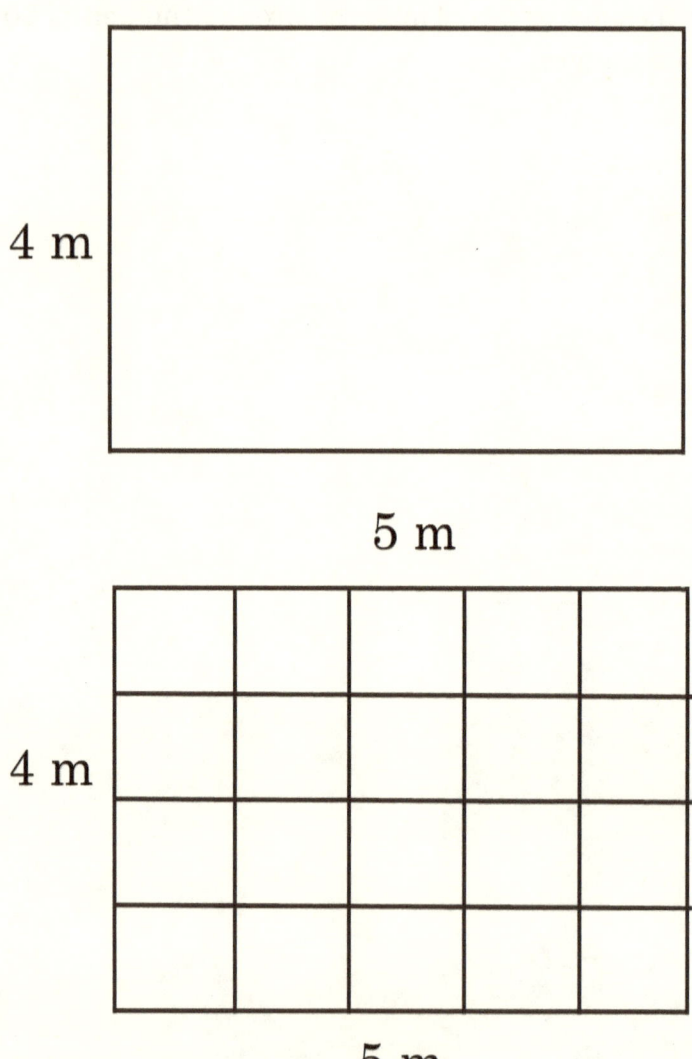

Fig. 42

Our friend Wolf MacStoat will build a wall for the pen. The wall will be 5 meters long and 2 meters high. He needs bricks to cover an area of 10 square meters. We obtain this area multiplying the base of the wall (5 meters) by its height (2 meters). We represent this result as

Area = base x height

$A = bh$

Fig. 43

Today we asked our friend Wolf MacStoat to build a pen with the following characteristics. Its area will be 200m^2 and one side should be the double of the other.

Since we do not know the dimensions of the sides, we call them X and the double of X (2X).

The formula for the area is A = bh

A = X(2X) = 2XX = 2X^2

X is measure in meters, X^2 in square meters.

Since the area is 200m^2, the equations to be solved is 2X^2 = 200

$2X^2 = 200$

If we divide by 2 one side, we have to divide by 2 the other.

$$\frac{2X^2}{2} = \frac{200}{2}$$

$X^2 = 100$

The answer is obtained by finding the number X, which multiplied by itself gives 100. The operation to discover such number is called square root.

$$\sqrt{X^2} = \sqrt{100}$$

$X = 10$

Therefore, the dimensions of the pen should be $X = 10m$ and $2X = 20m$

$A = bh = 10m(20m) = 200m^2$

Fig. 44

In general, how do we solve equations that include X^2?

The equations that involve areas include expressions of X^2 and they are called second grade equations or quadratic equations. There is a general second grade equation and all around the globe, they are solved with the general second grade formula or quadratic formula. In certain regions of Mexico, the quadratic formula is called "la chicharronera". Even the most serious and formal mathematics teacher, like Professor Andrés Jiménez Kaufmann, will say one day,

- We will solve this equation with "la chicharronera".

Quadratic equation:

$$aX^2 + bX + C = 0$$

La chicharronera:

$$X = \frac{-b \pm \sqrt{b^2 - 4ac}}{2a}$$

For example,

$$2X^2 - 2X - 12 = 0$$

$$a = 2 \quad b = -2 \quad c = -12$$

Fig. 45

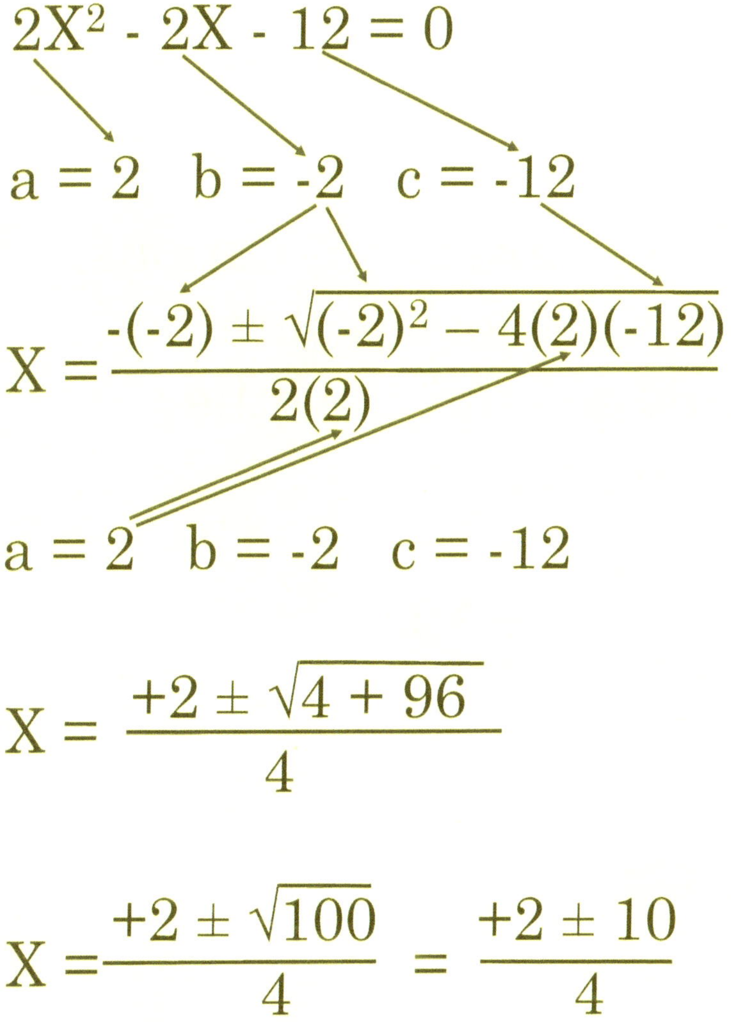

$$2X^2 - 2X - 12 = 0$$

$$a = 2 \quad b = -2 \quad c = -12$$

$$X = \frac{-(-2) \pm \sqrt{(-2)^2 - 4(2)(-12)}}{2(2)}$$

$$a = 2 \quad b = -2 \quad c = -12$$

$$X = \frac{+2 \pm \sqrt{4 + 96}}{4}$$

$$X = \frac{+2 \pm \sqrt{100}}{4} = \frac{+2 \pm 10}{4}$$

Fig. 46

$$X = \frac{+2 \pm 10}{4}$$

There are two solutions, one of them is obtained adding and the other subtracting.

$$X = \frac{+2 + 10}{4} = \frac{12}{4} = 3$$

$$X = \frac{+2 - 10}{4} = \frac{-8}{4} = -2$$

Fig. 47

Fig. 48

Now, let us solve some equations.

a. $2X^2 - 12X + 10$

b. $3X^2 - 18X + 15$

c. $5X^2 + 20X - 25$

d. $2X^2 + 8X - 10$

e. $2X^2 - 8X - 10$

f. $2X^2 + 12X + 10$

g. $3X^2 + 12X - 15$

h. $3X^2 - 12X - 15$

i. $3X^2 + 18X + 15$

j. $5X^2 - 30X + 25$

k. $5X^2 - 20X - 25$

l. $5X^2 + 30X + 25$

Answers. a) $X_1 = 1$, $X_2 = 5$; b) $X_1 = 1$, $X_2 = 5$; c) $X_1 = 1$, $X_2 = -5$; d) $X_1 = 1$, $X_2 = -5$; e) $X_1 = -1$, $X_2 = 5$; f) $X_1 = -1$, $X_2 = -5$; g) $X_1 = 1$, $X_2 = -5$; h) $X_1 = -1$, $X_2 = 5$; i) $X_1 = -1$, $X_2 = -5$; j) $X_1 = 1$, $X_2 = 5$; k) $X_1 = -1$, $X_2 = 5$; l) $X_1 = -1$, $X_2 = -5$.

Fig. 49

Most pieces of land are not rectangular in shape. They are predominately trapezoids. How can we calculate the area of a trapezoid? We do not know how to calculate the area of one trapezoid; however, we can easily calculate the added area of two trapezoids.

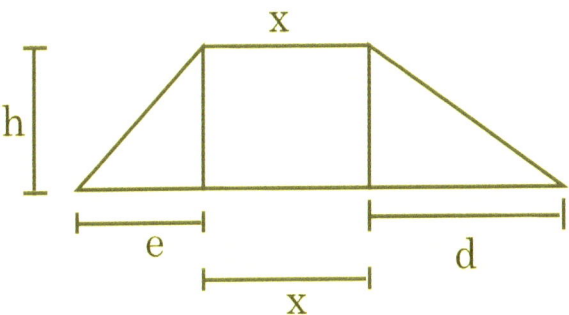

Instead of calculating the area of one trapezoid, we will calculate the areas of two trapezoids.

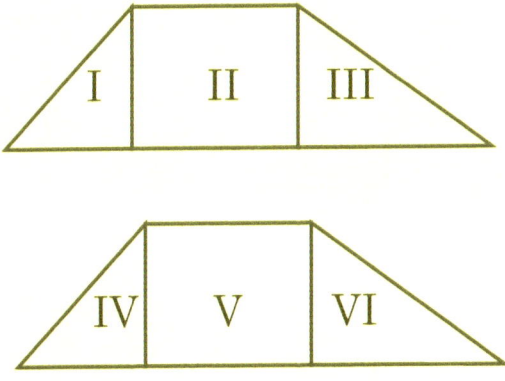

Fig. 50

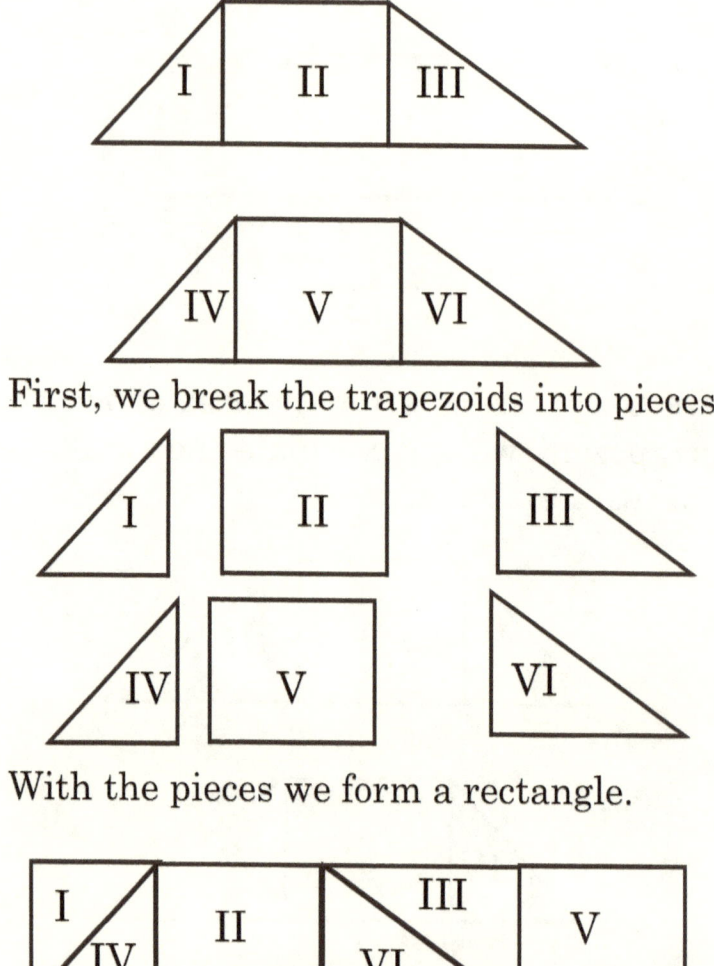

First, we break the trapezoids into pieces.

With the pieces we form a rectangle.

Fig. 51

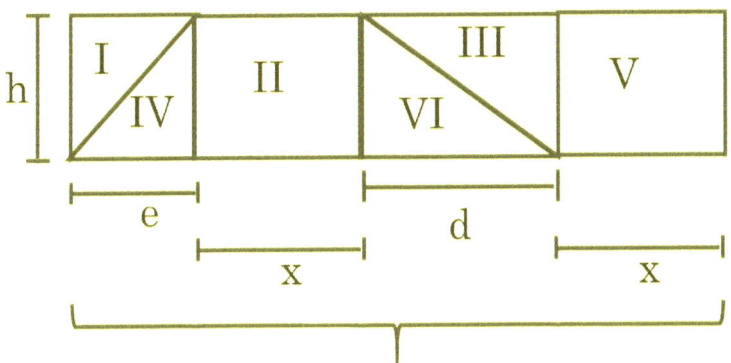

b = e + x + d + x = 2x + e + d

The area of the two trapezoids is the area of the rectangle = bh

= (e + x + d + x)h = (2x + e + d)h

The area of one trapezoid is half the area:

$$A = \frac{(2x + e + d)\,h}{2}$$

Fig. 52

Let us find the dimensions of a trapezoid whose area is 168m².

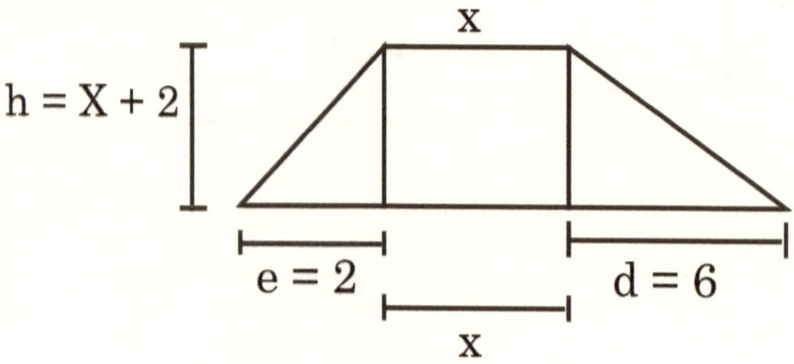

$$A = \frac{(2X + e + d)\, h}{2}$$

$$\frac{(2X + 8)\,(X + 2)}{2} = 168m^2$$

We multiply by 2 both sides of the equation. $(2X + 8)(X + 2) = 336$

With a few multiplications, this equation becomes:

$$2X^2 + 4X + 8X + 16 = 336$$

Fig. 53

$2X^2 + 4X + 8X + 16 = 336$

Which becomes

$2X^2 + 12X + 16 = 336$

$2X^2 + 12X + 16 - 336 = 336 - 336$

$2X^2 + 12X - 320 = 0$

We already know that we can use "la chicharronera" to solve this equation.

$$X = \frac{-b \pm \sqrt{b^2 - 4ac}}{2a}$$

$$X = \frac{-b \pm \sqrt{b^2 - 4ac}}{2a} = \frac{-12 \pm \sqrt{12^2 - 4(2)(-320)}}{2(2)}$$

$$X = \frac{-12 \pm \sqrt{144 - 4(2)(-320)}}{4} = \frac{-12 \pm \sqrt{144 + 2560}}{4}$$

$$X = \frac{-12 \pm \sqrt{2704}}{4} = \frac{-12 + 52}{4} = \frac{40}{4} = 10 \text{ meters}$$

Fig. 54

A few more problems will help us practice this new skill. Let us find the dimensions of the trapezoids whose areas are given.

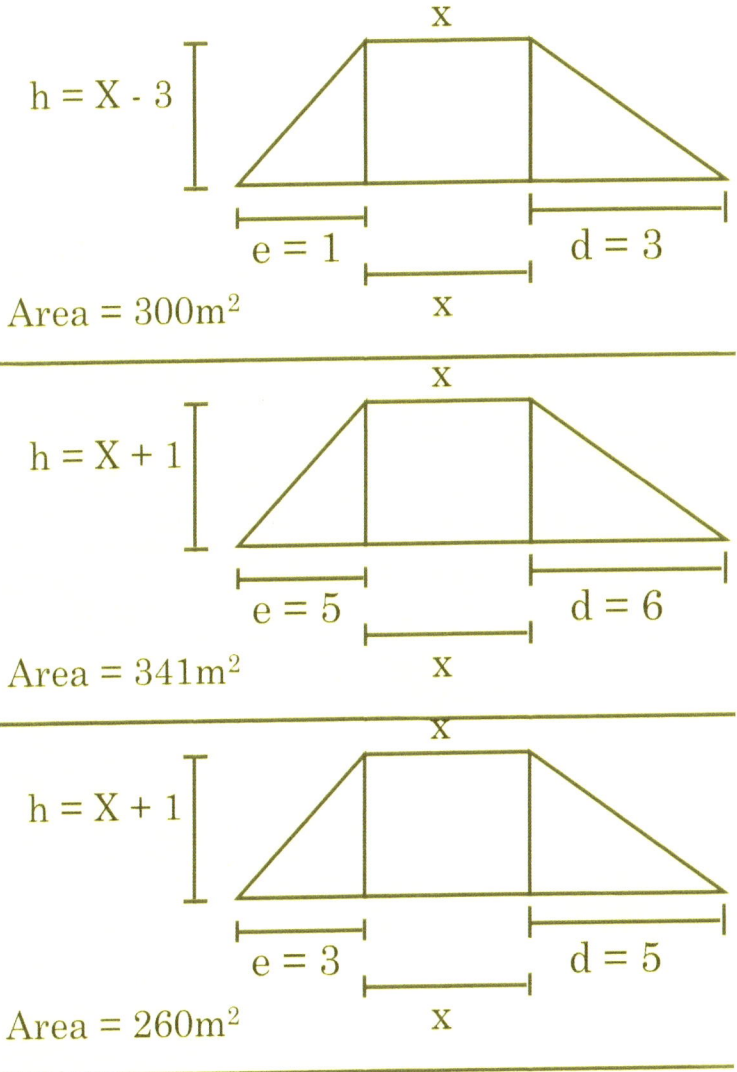

Fig. 55

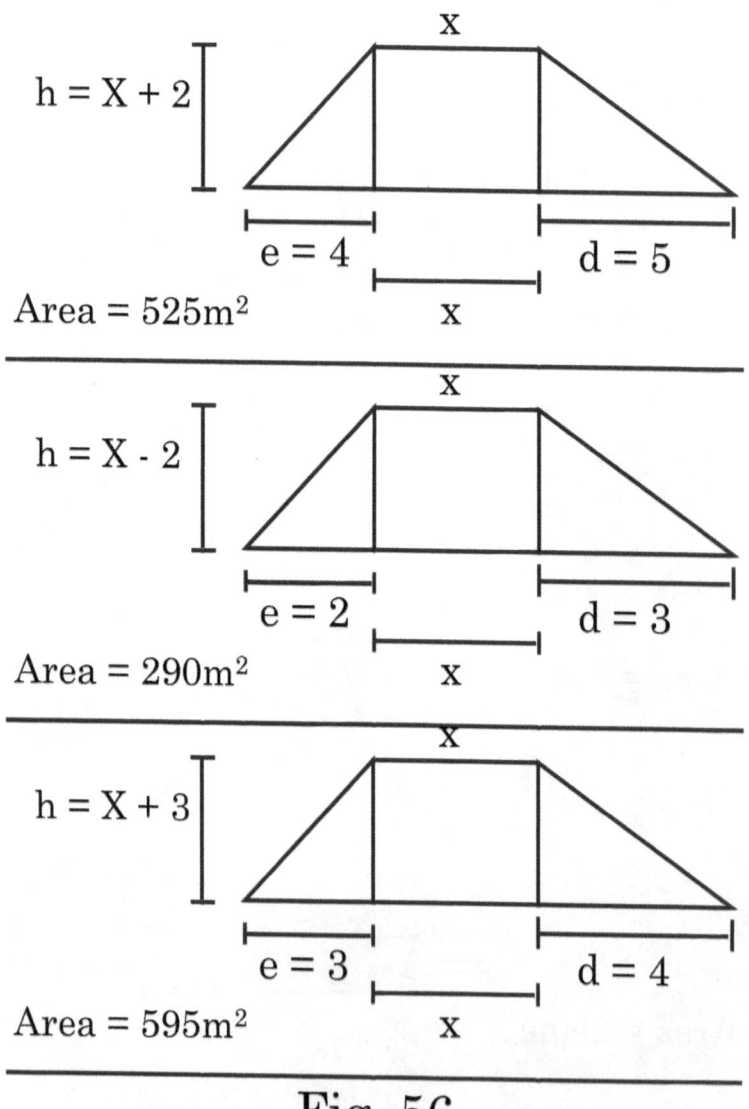

$$h = X + 2$$

$$e = 4$$

$$d = 5$$

$$\text{Area} = 525m^2$$

$$x$$

$$h = X - 2$$

$$e = 2$$

$$d = 3$$

$$\text{Area} = 290m^2$$

$$x$$

$$h = X + 3$$

$$e = 3$$

$$d = 4$$

$$\text{Area} = 595m^2$$

$$x$$

Fig. 56

Fig.57

PART V

A universe of space travelers

Fig. 58

11. Position and the exponential function

We, Charlotte and Jack, are writing from the deck of our spaceship, the Gilgamesh. Our systems failed and we are lost in this infinite vector space, that we call our universe. The distances in this universe are so vast that every second our position changes at an exponential rate in order to be able to cover the great lengths that we must travel.

People use the expression, "it increases exponentially", to mean that something grows very fast. How fast does the exponential function grow? Let us analyze a couple of examples.

Let us suppose that there is a single bacterium, and that it duplicates every hour.

Hours	Bacteria
0	1
1	2
2	4
3	8
4	16
5	32
6	64
7	128
8	256
9	512
10	1 thousand
11	2 thousand
12	4 thousand
13	8 thousand
14	16 thousand
15	32 thousand
16	64 thousand
17	128 thousand
18	256 thousand
19	512 thousand
20	1 million
21	2 million
22	4 million
23	8 million
24	16 million

Fig. 59

We started with one single bacterium duplicating every hour and the next day, we have more than 16 million bacteria. This is how fast the exponential function grows.

This example works fine to describe the dilemma we, Charlotte and Jack, are living in Gilgamesh, our spaceship. Our speed increases exponentially and the distance we travel duplicates every day.

Days	km
0	1
1	2
2	4
3	8
4	16
5	32
6	64
7	128
8	256
9	512
10	1 thousand
11	2 thousand
12	4 thousand
13	8 thousand
14	16 thousand
15	32 thousand
16	64 thousand
17	128 thousand
18	256 thousand
19	512 thousand
20	1 million
21	2 million
22	4 million
23	8 million
24	16 million
25	32 million
26	64 million
27	128 million
28	256 million
29	512 million
30	1 billion

Fig. 60

If we travel one km on the first day, by the end of the month we will have travelled more than one billion km. Thus, if we make a small mistake in the direction where we are heading, in one month we will be one billion km lost. To denote exponential changes, there is a certain notation that is used.

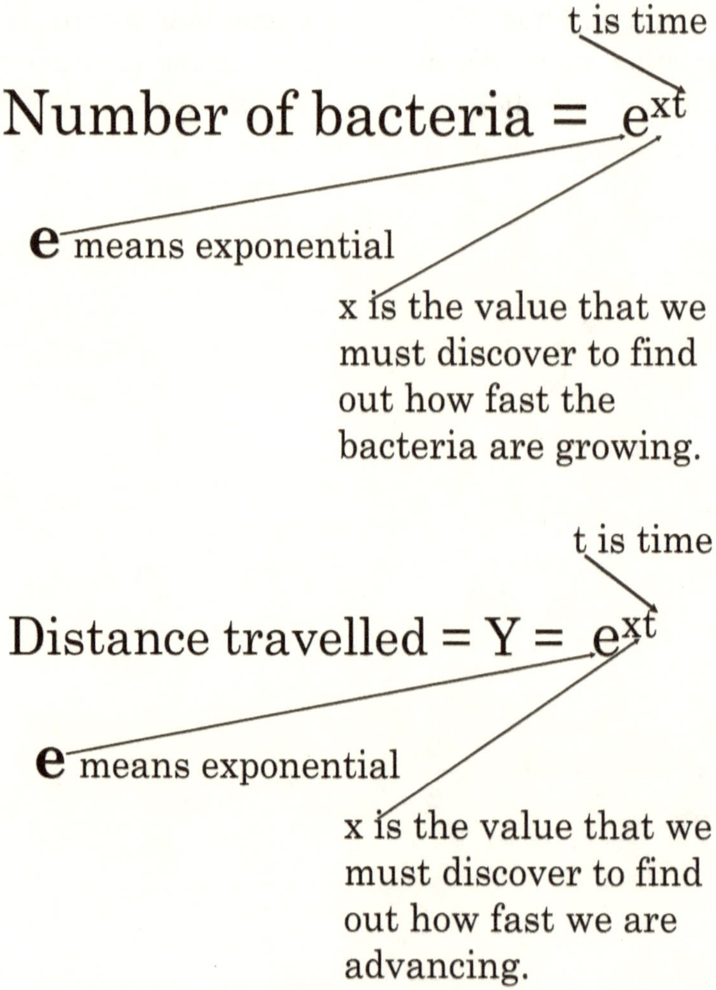

t is time

Number of bacteria = e^{xt}

e means exponential

x is the value that we must discover to find out how fast the bacteria are growing.

t is time

Distance travelled = Y = e^{xt}

e means exponential

x is the value that we must discover to find out how fast we are advancing.

Fig. 61

We will denote the distance travelled by Y.

For us to know our exact position in the space vector that we call our universe, we need to find the value of x in the equation:

$Y = e^{xt}$

Our position is always changing. Every new value of t (the time) gives us a new value of Y (our position). How fast our position changes is called our speed. The mathematical notation for speed is:

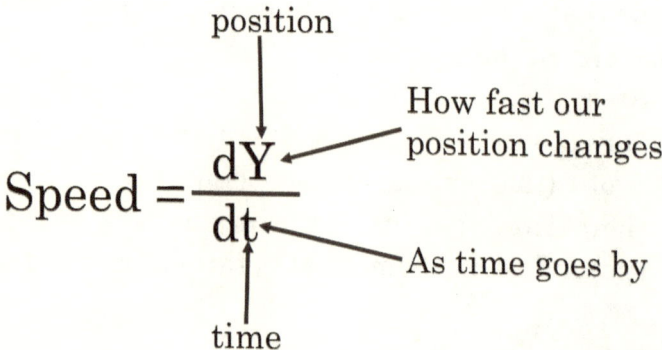

Speed measures how fast our position changes as time goes by.

Not only our position is changing, our speed is also changing. It is just as when you go from 70 km/h to 80 km/h. The speed increases.

The acceleration describes how the speed changes.

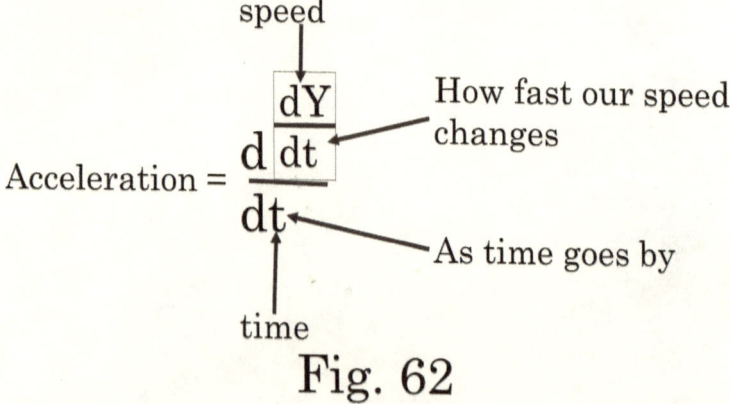

Fig. 62

The mathematical notation for acceleration is:

$$\text{Acceleration} = \frac{d^2Y}{dt^2}$$

How fast our speed changes

As time goes by

time

$$\text{Position} = Y$$

$$\text{Speed} = \frac{dY}{dt}$$

$$\text{Acceleration} = \frac{d^2Y}{dt^2}$$

Fig. 63

We are lost in space. We would like to know our position (Y). We know that our position changes exponentially as time goes by.

$Y = e^{xt}$

However, we cannot know our whereabouts because we ignore the value of x. Nevertheless, we do have some information. There is an equation that will help us find our position in space. Before breaking down, our computer delivered us an equation that says that twice the acceleration minus six times our speed, minus 20 times our position is zero.

Twice the acceleration,

$$2\,\frac{d^2Y}{dt^2} - 6\,\frac{dY}{dt} - 20Y = 0$$

minus six times our speed, minus 20 times our position is zero.

With that equation, how can we find the value of x in $Y = e^{xt}$?

In our universe we cannot. However, we have a transformation device that changes this equations into a three number vector.

$$2\,\frac{d^2Y}{dt^2} - 6\,\frac{dY}{dt} - 20Y = 0$$

$$(2, -6, -20)$$

Fig. 64

Fig. 65

$$2\frac{d^2Y}{dt^2} - 6\frac{dY}{dt} - 20Y = 0$$

$$(2, -6, -20)$$

And we send this vector to a parallel universe where we are farmers and we own mammoths. In that universe, these three numbers become a different equation:

$$(2, -6, -20)$$

$$2X^2 - 6X - 20 = 0$$

In that universe we have a formula to sove this equation,
"la chicharronera".

$$X = \frac{-b \pm \sqrt{b^2 - 4ac}}{2a} = \frac{6 \pm \sqrt{6^2 - 4(2)(-20)}}{2(2)}$$

$X_1 = 5$ and $X_2 = -2$

Fig. 66

Now, we bring the results ($X_1 = 5$ and $X_2 = -2$) back to us and substitute them in the formula for position $Y = e^{xt}$.

Finally we know where we are. Our exact position is:

$$Y = e^{5t} + e^{-2t}$$

This procedure is called solving a differential equation. It involves travelling from one universe, where we do not know how to solve a differential equation, to another where we can use "la chicharronera" to solve another equation.

Fig. 67

Fig. 68

Now, let us solve some differential equations.

Twice the acceleration,

$$2\,\frac{d^2Y}{dt^2} - 12\,\frac{dY}{dt} + 10Y = 0$$

minus 12 times our speed, plus 10 times our position is zero.

Thrice the acceleration,

$$3\,\frac{d^2Y}{dt^2} - 18\,\frac{dY}{dt} + 15Y = 0$$

minus 18 times our speed, plus 15 times our position is zero.

Five times the acceleration,

$$5\,\frac{d^2Y}{dt^2} + 20\,\frac{dY}{dt} - 25Y = 0$$

plus 20 times our speed, minus 25 times our position is zero.

Fig. 69

Twice the acceleration,

$$2\,\frac{d^2Y}{dt^2} - 8\,\frac{dY}{dt} - 10Y = 0$$

minus eight times the speed, minus 10 times our position is zero.

Twice the acceleration,

$$2\,\frac{d^2Y}{dt^2} +12\,\frac{dY}{dt} + 10Y = 0$$

plus 12 times our speed, plus 10 times our position is zero.

Thrice the acceleration,

$$3\,\frac{d^2Y}{dt^2} +12\,\frac{dY}{dt} - 15Y = 0$$

plus 12 times our speed, minus 15 times our position is zero.

Fig. 70

Fig. 71

12. Circular and elliptical orbits

Planets, moons, space stations, communication satellites, and space cities orbit around larger space bodies that attract them with a gravitational force. Such orbits have a defined closed shape, mostly elliptical, occasionally circular. Both circular and elliptical orbits are described by a two variable quadratic equation.

In the case of the circle, the equation is

$(x - h)^2 + (y - k)^2 = r^2$

Where the center is (h, k) and the radius is r.

Let us look at an example.

134

Pay attention to the sign, it is the opposite.

$$(x - 2)^2 + (y - 1)^2 = 3^2$$

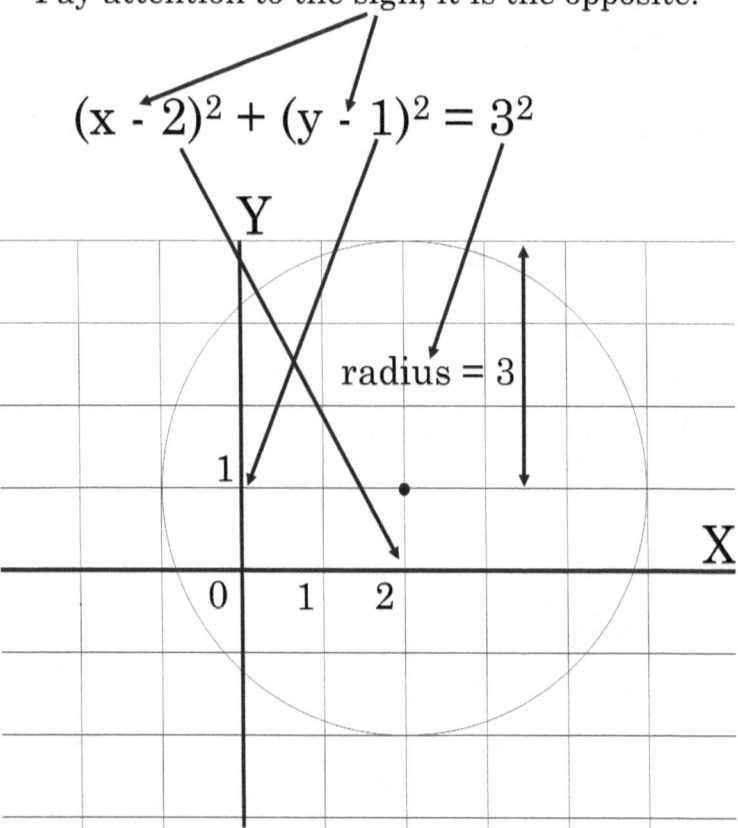

Fig. 72

Pay attention to the sign, it is the opposite.

$$(x + 2)^2 + (y + 1)^2 = 3^2$$

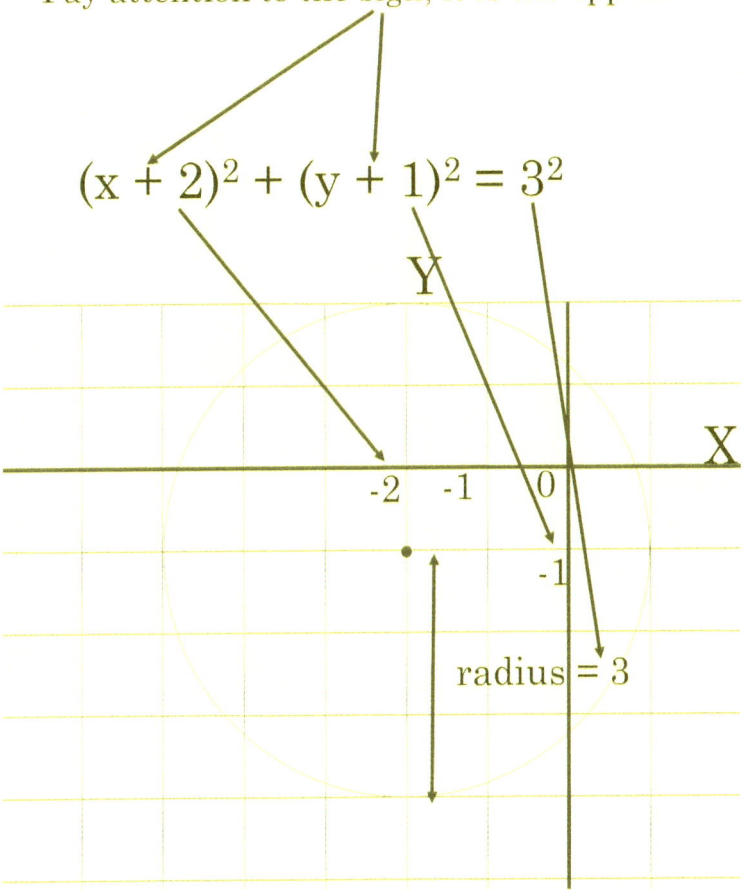

Fig. 73

Fig. 74

Pay attention to the sign, it is the opposite.

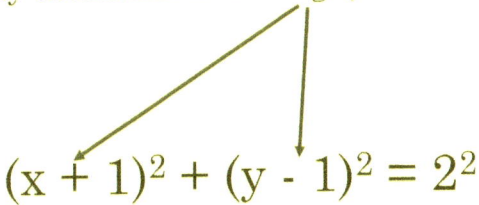

$$(x + 1)^2 + (y - 1)^2 = 2^2$$

Fig. 75

Pay attention to the sign, it is the opposite.

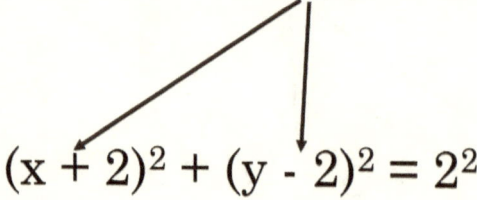

$$(x + 2)^2 + (y - 2)^2 = 2^2$$

Y

2

1

-2   -1   0   X

Fig. 76

Pay attention to the sign, it is the opposite.

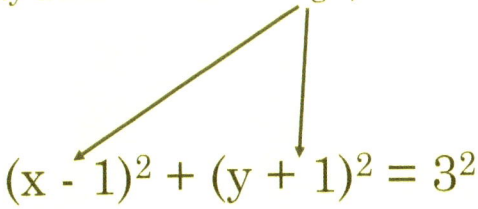

$$(x - 1)^2 + (y + 1)^2 = 3^2$$

Fig. 77

Fig. 78

The equation of the ellipse is:

$$\frac{(x - h)^2}{a^2} + \frac{(y - k)^2}{b^2} = 1$$

(h, k) is the center of the ellipse, a is the semiaxis in the direction of X, and b is the semiaxis in the direction of Y.

$$\frac{(x - 2)^2}{2^2} + \frac{(y - 1)^2}{3^2} = 1$$

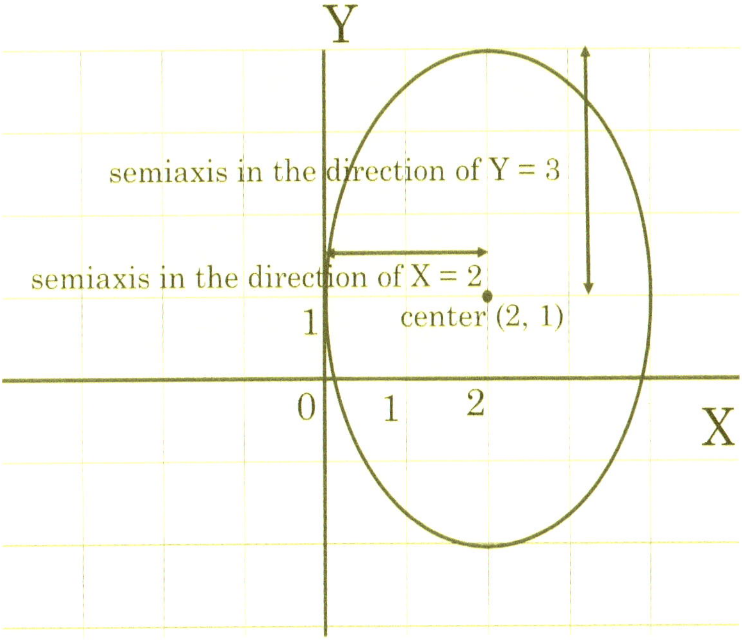

Y

semiaxis in the direction of Y = 3

semiaxis in the direction of X = 2

center (2, 1)

1

0   1   2

X

Fig. 79

Fig. 80

$$\frac{(x - 1)^2}{1^2} + \frac{(y - 3)^2}{3^2} = 1$$

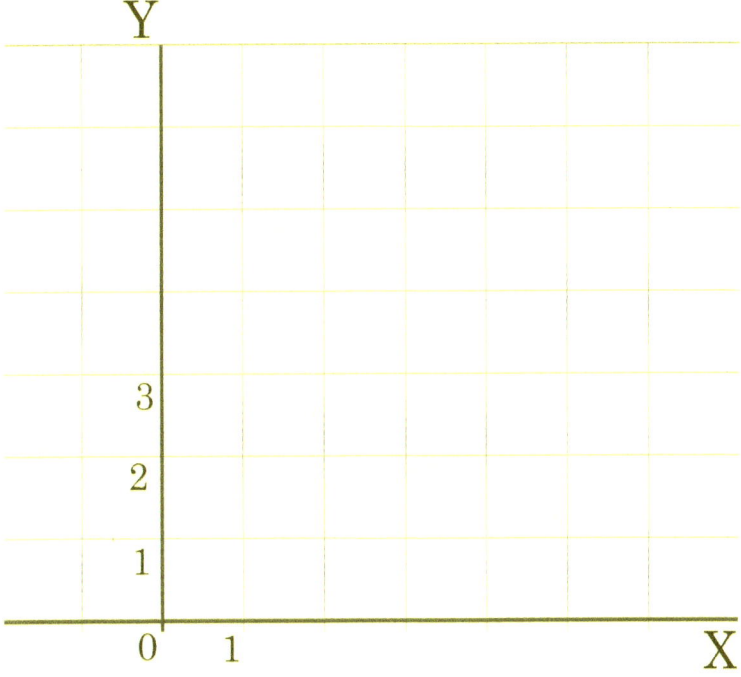

Fig. 81

$$\frac{(x + 2)^2}{3^2} + \frac{(y - 1)^2}{2^2} = 1$$

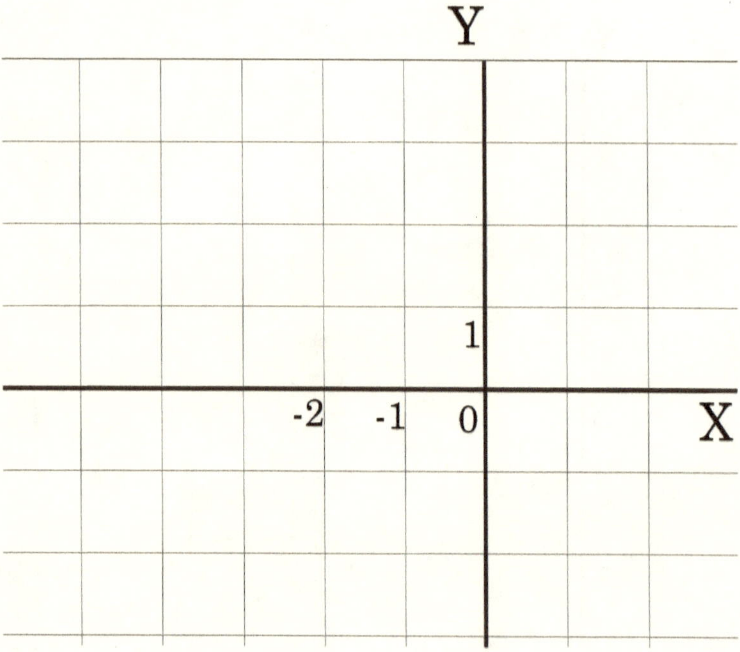

Fig. 82

$$\frac{(x + 2)^2}{2^2} + \frac{(y - 1)^2}{3^2} = 1$$

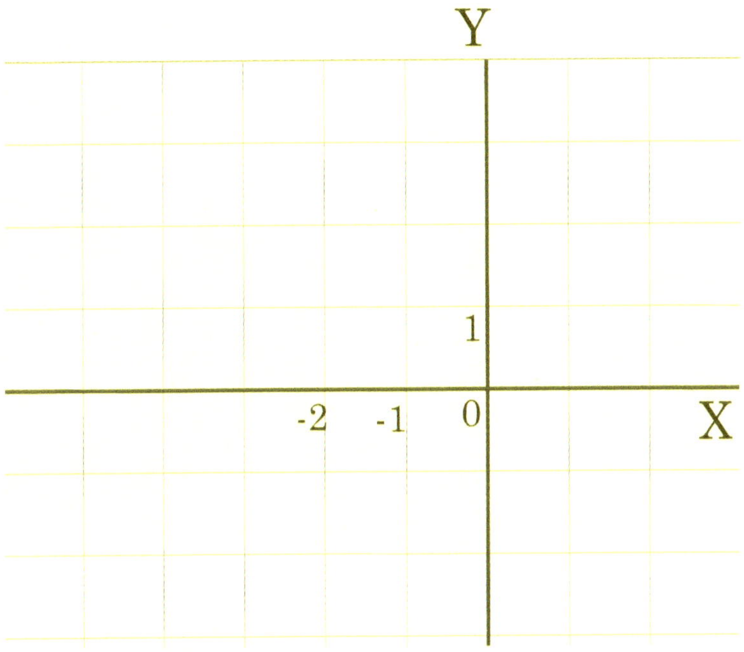

Fig. 83

EPILOGUE

What is this book really about?

This book is about how to transform a differential equation in a universe, into a quadratic equation in another vector space. By the way, the device to perform such a transformation is called an isomorphism. If you want to cause a very favorable impression on someone, you should only say, "to solve a differential equation of this type, you transform it into a quadratic equation by means of an isomorphism of vector spaces". It sound very nerdy, indeed.

A transformation to travel to a parallel universe

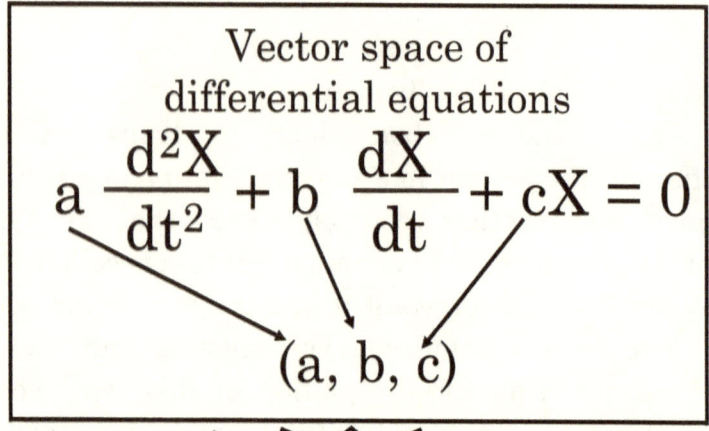

Vector space of
differential equations

$$a\,\frac{d^2X}{dt^2} + b\,\frac{dX}{dt} + cX = 0$$

$$(a, b, c)$$

Isomorphism of
vector spaces

$$(a, b, c)$$

$$ax^2 + bx + c = 0$$

Vector space of quadratic equations

Fig. 84

Nevertheless, the book is really about you and its purpose is to show you that quadratic equations are not meaningless.

First, we learned a few tools to be able to deal with quadratic equations. Then we studied how quadratic can be used to solve problems related to the areas of trapezoids. Later on, you learned how to solve some differential equations in exactly the same way as engineering students solve them at college. Finally, we saw that from a quadratic expression you can retrieve information to draw a circle or an ellipse. There are hundreds of applications of quadratic equations in all fields of knowledge.

$$\frac{d^2X}{dt^2} - \frac{dX}{dt} - 6X = 0$$

$$\frac{(X-1)^2}{3^2} + \frac{(Y-2)^2}{4^2} = 1$$

Fig. 85

I hope that as you read this book, you learned something about quadratic equations and that from now on, whenever you find one, you will see it as a friend and not as a foe. I wish you the best of luck and success in your life and in your future mathematical endeavors.